Farm Animals Coloring
Book for Kids

By: Tomeu D.F.

this book belongs to:

How to use this book

Welcome to "Farm Animals Coloring Book for Kids"! This book is designed for children to have fun coloring beautiful animals that you would find in a farm from around the world, while learning about them. Each page presents a creative and educational challenge.

- On the left pages, you will encounter a black background with a "Did you know?" information of each animal in white color text.

- On the right pages, you will encounter the illustration for coloring.

- Don't forget to use the "tip color" that the left page gives to you and unleash the artist within you!
- Learn and enjoy coloring that 40 beauty farm animals!

From the author:

Dive into the delightful world of farm animals with this second book of the series, "Learn and Color with Wonders of Nature". Each carefully crafted illustration invites you to explore the charm and diversity of our favorite barnyard companions.

Discover the joy of coloring as you bring these farm animals to life, and let your imagination run free. This book is not just about art; it's a fun and educational journey into the world of agriculture animals.

Celebrate the beauty of these creatures and enjoy a creative escape.

Tomeu D.F.

I value your opinion and would appreciate it if you could share it with me and other customers. If you have a moment, I would be grateful if you could leave a review on the book's Amazon page. Thank you very much for your support.

Color Test Page

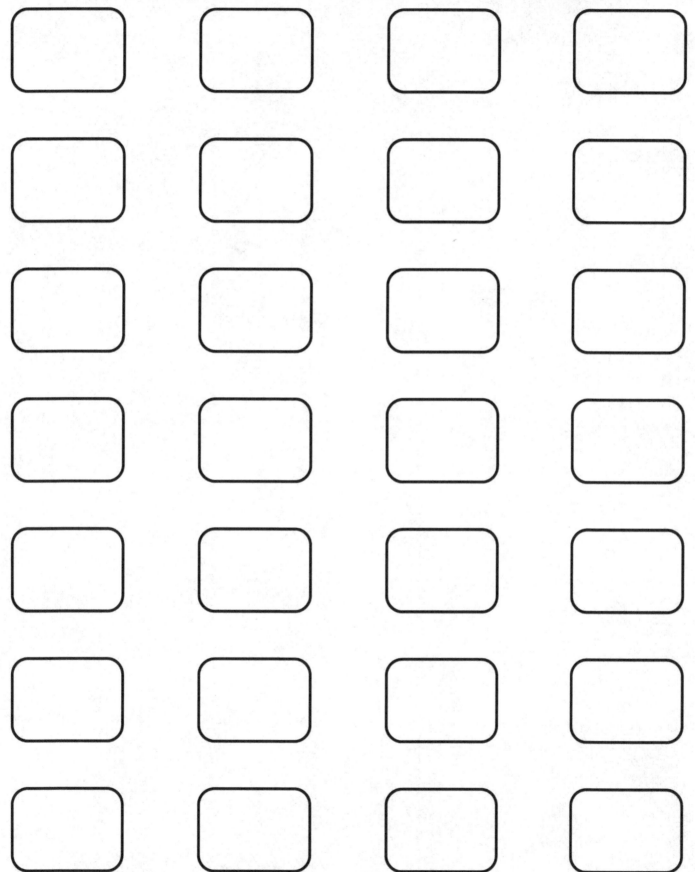

Farm Animals Coloring Book For Kids

Take your FREE gift!

As a token of appreciation for purchasing this book, follow the QR for a gift you won't want to miss.

Two FREE coloring books:

Or... send an email to
gifts@tomeudf.com
with the word:
'FARMGIFT'

COW

- Class: Mammalia
- Order: Artiodactyla
- Family: Bovidae
- Origin: Worldwide, originally descended from wild oxen.

DID YOU KNOW?

Cows are smart and social beings. In their herds, they form close friendships and have a complex social structure. Cows communicate using various moos, each with a unique purpose, from calling their calves to expressing contentment. Their intelligence and emotional depth make them more than just grazers; they're wonderful and socially savvy creatures!

COLOR TIP

For the cow's fur, use a mix of warm browns or blacks and white. Their coats can vary, so feel free to add lighter and darker shades to capture their natural diversity.

DONKEY

- Class: Mammalia
- Order: Perissodactyla
- Family: Equidae
- Origin: Worldwide, descended from Africa.

DID YOU KNOW?

Donkeys are not just hardworking farm pals; they're also clever thinkers. With an impressive memory, they can remember places and friends for a surprisingly long time. Their sharp minds and loyalty make them not only valuable helpers in various tasks but also endearing companions on the farm.

COLOR TIP

For the donkey's fur, use shades of gray and brown. Their coats can vary, but these earthy tones capture the essence of these hardworking and friendly creatures.

OSTRICH

- Class: Birds
- Order: Struthioniformes
- Family: Struthionidae
- Origin: Africa

DID YOU KNOW?

Ostriches are the world's largest and heaviest birds. Despite their inability to fly, they are exceptional runners, capable of reaching speeds up to 45 miles per hour. Additionally, ostriches have powerful kicks that can be used for self-defense.

COLOR TIP

For the ostrich's plumage, use shades of brown and gray. their feathers are adaptive and help them blend into their natural surroundings, showcasing nature's perfect camouflage.

HORSE

- Class: Mammalia
- Order: Perissodactyla
- Family: Equidae
- Origin: Worldwide, descended from wild ancestors.

DID YOU KNOW?

Horses have been companions to humans for thousands of years. These majestic creatures not only provide transportation but also form strong emotional bonds with their riders. They have a remarkable ability to understand human emotions and respond to subtle cues, making them exceptional partners in various activities.

COLOR TIP

For the horse's coat and mane, use a mix of browns and blacks. Their colors can vary greatly, so feel free to add different shades to capture the beauty of these magnificent animals.

GOAT

- Class: Mammalia
- Order: Artiodactyla
- Family: Bovidae
- Origin: Worldwide, with various breeds.

DID YOU KNOW?

Goats are curious and clever creatures. They have rectangular pupils, giving them a wide field of vision to detect predators. Goats are also known for their diverse vocalizations, from gentle bleats to robust calls, and they're excellent climbers, navigating rocky terrain with ease.

COLOR TIP

For the goat's fur, use a mix of whites, browns, and blacks. Goats come in various colors and patterns, so get creative to capture the charming diversity of these farm friends.

CAMEL

- Class: Mammalia
- Order: Artiodactyla
- Family: Camelidae
- Origin: Various regions, native to the Middle East.

DID YOU KNOW?

Camels are the "ships of the desert," perfectly adapted to arid environments. They have unique features like a hump for storing fat, enabling them to endure long journeys without water. Camels are also known for their gentle and social nature, forming close bonds with their human companions.

COLOR TIP

For the camel's fur, use shades of brown and tan. Their coats are designed to blend with the sandy desert surroundings, making them resilient travelers in their arid homes.

PIG

- Class: Mammalia
- Order: Artiodactyla
- Family: Suidae
- Origin: Worldwide, descended from wild boars.

DID YOU KNOW?

Pigs are incredibly intelligent animals, often compared to dogs in cognitive abilities. They are social creatures that enjoy rooting around in the soil with their snouts. Pigs also communicate through a range of oinks, grunts, and squeals, expressing their feelings and needs.

COLOR TIP

For the pig's skin, use various shades of pink. Pigs come in different colors, but the classic pink hue captures their delightful and friendly appearance.

BEE

- Class: Insecta
- Order: Hymenoptera
- Family: Apidae
- Origin: WorldWide

DID YOU KNOW?

Bees are like tiny gardeners with a sweet tooth, and they're also talented chefs! They buzz around collecting nectar from flowers, turning it into delicious honey. Bees live in busy colonies and communicate through dances. they're not just hard workers; they're vital pollinators that contribute to the beauty of nature.

COLOR TIP

For the bee's body, use bright browns and blacks. Capture the vibrant and bold appearance of these essential pollinators, adding patterns to showcase the distinct and energetic nature of bees.

HEN

- Class: Birds
- Order: Galliformes
- Family: Phasianidae
- Origin: Worldwide

DID YOU KNOW?

Hens are amazing multitaskers. Not only do they lay eggs, but they're also excellent insect hunters, helping to keep the farmyard bug-free. Hens are social birds that enjoy scratching the soil in search of tasty treats, and they even have unique clucking sounds to communicate with their chicks.

COLOR TIP

For the hen's feathers, use a mix of browns and reds. Hens come in various breeds and colors, so you can get creative to capture the diverse and charming appearances of these feathered friends.

SHEEP

- Class: Mammalia
- Order: Artiodactyla
- Family: Bovidae
- Origin: Worldwide

DID YOU KNOW?

Sheep are not just woolly wonders; they are also great landscapers. Their love for grazing helps maintain grassy fields and prevents overgrowth. Sheep have excellent memories and can recognize faces, including their fellow flock members. They're social animals that prefer safety in numbers.

COLOR TIP

For the sheep's wool, use various shades of white. While sheep can come in different colors, the classic white or light grey captures the essence of their fluffy and cozy appearance.

PHEASANT

- Class: Birds
- Order: Galliformes
- Family: Phasianidae
- Origin: Asia

DID YOU KNOW?

Pheasants are renowned for their vibrant plumage, especially the males. Their striking colors help them attract mates during courtship displays. Pheasants are skilled fliers and prefer to run or walk rather than fly, but when they take flight, their wings produce a distinctive whirring sound.

COLOR TIP

For the pheasant's feathers, use a mix of bold and bright colors. Capture the essence of their stunning appearance, with hues like greens, reds, and golds to make them stand out in the fields.

PYGMY GOAT

- Class: Mammalia
- Order: Artiodactyla
- Family: Bovidae
- Origin: Various regions, bred from wild goats.

DID YOU KNOW?

Pygmy goats may be small in size, but they're big in personality. These playful and curious goats are known for their friendly nature and love to climb. Despite their tiny stature, pygmy goats are excellent jumpers and can clear obstacles with surprising ease.

COLOR TIP

For the pygmy goat's coat, use a mix of whites, browns, and blacks. Their charming colors make them irresistible, so feel free to add different shades to capture the adorable diversity of these miniature farm friends.

GOOSE

- Class: Birds
- Order: Anseriformes
- Family: Anatidae
- Origin: Worldwide

DID YOU KNOW?

Geese are not only excellent navigators during migration but also vigilant guardians of their territory. They're known for their loud honks, which serve as warning calls. Geese are highly social birds that form strong family bonds and often mate for life.

COLOR TIP

For the goose's feathers, use shades of whites, grays, and browns. These colors reflect the natural tones of both domestic and wild geese, capturing their elegant and distinctive appearance.

TURKEY

- Class: Birds
- Order: Galliformes
- Family: Phasianidae
- Origin: North America

DID YOU KNOW?

Turkeys are famous for their elaborate fan-shaped tails, especially during courtship displays. Contrary to popular belief, they are surprisingly agile flyers and roost in trees at night for safety. Turkeys also have a range of vocalizations, from gobbling to purring, to communicate with each other.

COLOR TIP

For the turkey's feathers, use a mix of earthy tones such as browns, blacks, and iridescent hues. Capture the beauty of their plumage, showcasing the diverse colors that make turkeys stand out in the wild.

CARP

- Class: Actinopterygii
- Order: Cypriniformes
- Family: Cyprinidae
- Origin: Asia

DID YOU KNOW?

Carp are resilient and adaptable fish known for their ability to thrive in various aquatic environments. They have a keen sense of smell and can detect scents in the water, helping them locate food. Carp also exhibit diverse coloration, with some varieties displaying vibrant scales.

COLOR TIP

For the carp's scales, use a mix of golds, silvers, and oranges. Capture the shimmering and reflective qualities of their scales, bringing out the vibrant colors that make carp fascinating aquatic residents.

GUINEA PIG

- Class: Mammalia
- Order: Rodentia
- Family: Caviidae
- Origin: South America

DID YOU KNOW?

Guinea pigs are social and affectionate rodents that form strong bonds with their human companions. they communicate through various sounds, including squeaks, purrs, and chirps. Despite their name, guinea pigs are not pigs at all; they are small rodents with a gentle and friendly disposition.

COLOR TIP

For the guinea pig's fur, use a mix of browns, whites, and blacks. these adorable pets come in various coat patterns, so have fun adding different shades to capture the charming diversity of guinea pigs.

SNAIL

- Class: Gastropoda
- Order: Stylommatophora
- Family: Helicidae
- Origin: Worldwide

DID YOU KNOW?

Snails are slow-moving but fascinating creatures. they carry their homes on their backs, and their spiral-shaped shells come in various colors and patterns. Snails move by gliding on a trail of slime, leaving a unique and intricate path behind them.

COLOR TIP

For the snail's shell, use a mix of soft browns, creams, and subtle patterns. Add a touch of creativity to showcase the beautiful and varied shells that make each snail unique.

DUCK

- Class: Birds
- Order: Anseriformes
- Family: Anatidae
- Origin: Worldwide

DID YOU KNOW?

Ducks are not just graceful swimmers; they are excellent fliers too. They use their waterproof feathers to stay buoyant in the water and have a range of quacks, whistles, and coos to communicate with each other. Ducks are also known for their love of foraging, dabbling, and tipping upside-down to find tasty treats.

COLOR TIP

For the duck's feathers, use a mix of earthy tones such as browns, greens, and yellows. These colors capture the natural beauty of ducks, whether they're gliding on water or waddling on land.

LLAMA

- Class: Mammalia
- Order: Artiodactyla
- Family: Camelidae
- Origin: South America

DID YOU KNOW?

Llamas are known for their gentle nature and are often used as pack animals in mountainous terrain. They have a unique hum that serves as a form of communication, and they are excellent at navigating rocky landscapes. Llamas are also prized for their soft and warm wool.

COLOR TIP

For the llama's fur, use shades of white, brown, and gray. Their coats can have a delightful mix of colors, so feel free to add different tones to capture the charming diversity of llamas.

PIGEON

- Class: Birds
- Order: Columbiformes
- Family: Columbidae
- Origin: Worldwide

DID YOU KNOW?

Pigeons are intelligent birds with an incredible homing instinct. They have been used throughout history to carry messages over long distances. Pigeons also exhibit a gentle and calm demeanor, and they come in a variety of colors and patterns.

COLOR TIP

For the pigeon's feathers, use a mix of grays, blues, and subtle iridescence. Capture the elegant and muted tones that make pigeons distinctive urban birds.

CALF

- Class: Mammalia
- Order: Artiodactyla
- Family: Bovidae
- Origin: Worldwide, descended from cows.

DID YOU KNOW?

Calves are playful and curious young cattle. They form strong bonds with their mothers and enjoy exploring their surroundings. Calves are quick learners, and as they grow, they develop unique personalities. They're a symbol of new beginnings on the farm.

COLOR TIP

For the calf's fur, use a mix of light browns and whites. Capture the soft and fuzzy appearance of these adorable farm youngsters, adding warmth to their delightful presence.

ROOSTER

- Class: Birds
- Order: Galliformes
- Family: Phasianidae
- Origin: Worldwide

DID YOU KNOW?

Roosters are the kings of the barnyard, known for their crowing at sunrise. They have vibrant plumage and distinctive combs, and their crowing serves not only as a wake-up call but also as a way to establish territory. Roosters are also watchful protectors, keeping an eye on their flock.

COLOR TIP

For the rooster's feathers, use bold and bright colors. Capture the striking and vivid hues that make roosters stand out, with a mix of reds, greens, and golds to showcase their regal appearance.

BORDER COLLIE

- Class: Mammalia
- Order: Carnivora
- Family: Canidae
- Origin: Scotland and England.

DID YOU KNOW?

Border Collies are incredibly intelligent and agile herding dogs. They are known for their remarkable ability to understand and follow commands, making them excellent working dogs. Border Collies have boundless energy and a strong work ethic, and they form deep bonds with their human companions.

COLOR TIP

For the Border Collie's fur, use a mix of blacks, whites, and browns. Capture the sleek and agile appearance of these brilliant dogs, adding depth to their coat to highlight their intelligent and friendly nature.

BULL

- Class: Mammalia
- Order: Artiodactyla
- Family: Bovidae
- Origin: Worldwide

DID YOU KNOW?

Bulls are powerful and iconic symbols of strength. While they are often associated with aggression, they are also gentle and can form close bonds with humans. Bulls play a vital role in agriculture, contributing to breeding programs and ensuring the health of livestock herds.

COLOR TIP

For the bull's coat, use a mix of deep browns and blacks. Highlight the powerful and majestic presence of bulls with rich and bold colors, capturing the essence of these magnificent creatures.

TORTOISE

- Class: Reptilia
- Order: Testudines
- Family: Testudinidae
- Origin: Various regions

DID YOU KNOW?

Land turtles, or tortoises, are known for their slow and steady pace. They have a remarkable ability to retract into their protective shells when threatened. Tortoises are herbivores and enjoy a diet rich in vegetation. Their longevity is impressive, with some species living well over a century.

COLOR TIP

For the land turtle's shell, use a mix of earthy tones such as browns, grays and greens. Add patterns and textures to capture the rugged and unique appearance of these wise and ancient reptiles.

TURTLEDOVE

- Class: Birds
- Order: Columbiformes
- Family: Columbidae
- Origin: Worldwide

DID YOU KNOW?

turtledoves are symbols of love and devotion. they make sweet cooing sounds, and they're really good at staying with one special friend for life. turtledoves love going on adventures, flying to new places when it's time for a vacation. their feathers are like soft blankets, making them look extra fancy and calm.

COLOR TIP

For the turtledove's feathers, use soft and muted colors like grays, whites, and light browns. Capture the gentle and calming appearance of these beautiful birds, adding subtle tones to showcase their elegance.

PONY

- Class: Mammalia
- Order: Perissodactyla
- Family: Equidae
- Origin: Worldwide

DID YOU KNOW?

Ponies are like pint-sized horses, but don't let their small stature fool you—they're full of personality! Known for their friendly nature and playful antics, ponies make great companions. They come in various colors and patterns, and their fluffy manes and tails add an extra touch of cuteness.

COLOR TIP

For the pony's coat, use a mix of browns, blacks, and whites. Get creative with different shades to capture the adorable and diverse appearances of these delightful little equines.

FROG

- Class: Amphibia
- Order: Anura
- Family: Various families, including Ranidae and Hylidae
- Origin: Worldwide

DID YOU KNOW?

Frogs are fantastic jumpers and singers of the pond symphony! They start as tiny tadpoles and transform into hoppers. Frogs have smooth, moist skin and big eyes to see bugs for dinner. Some even change colors! Next time you hear a ribbit, you'll know a frog is saying hello.

COLOR TIP

For the frog's skin, use shades of greens and browns. Frogs come in different colors, so feel free to add a mix of tones to capture the vibrant and lively spirit of these amphibious friends.

REINDEER

- Class: Mammalia
- Order: Artiodactyla
- Family: Cervidae
- Origin: Arctic and Subarctic regions

DID YOU KNOW?

Reindeer are like Santa's helpers in the snowy North! they have special hooves for digging through the snow to find food. Both males and females grow antlers, and they're great at pulling sleighs. In winter, their fur turns super thick, like a warm and cozy sweater!

COLOR TIP

For the reindeer's fur, use shades of browns, grays, and whites. Capture the warmth of their winter coat by adding layers of different tones, creating a fluffy and snug appearance

SALMON

- Class: Actinopterygii
- Order: Salmoniformes
- Family: Salmonidae
- Origin: North Atlantic and Pacific oceans

DID YOU KNOW?

Salmon are incredible swimmers, leaping upstream to lay their eggs in cool, freshwater rivers. They start as tiny eggs, become playful fry, and then turn into mighty swimmers. Salmon have a keen sense of smell, helping them find their way back to their birthplace for the next generation.

COLOR TIP

For the salmon's scales, use shades of silver, pinks, and blues. Capture the shimmering and vibrant colors that make salmon stand out, especially during their exciting journey upstream.

MOLE

- Class: Mammalia
- Order: Eulipotyphla
- Family: Talpidae
- Origin: Worldwide

DID YOU KNOW?

Moles are amazing diggers with strong front paws made for tunneling. They live most of their lives underground, where they create intricate tunnel systems. Moles have velvety fur and tiny eyes, but their keen sense of touch helps them navigate through dark tunnels to find tasty insects.

COLOR TIP

For the mole's fur, use shades of grays and browns. Capture the soft and velvety appearance of their fur, adding subtle textures to bring out the charming features of these subterranean experts.

PEACOCK

- Class: Birds
- Order: Galliformes
- Family: Phasianidae
- Origin: Asia

DID YOU KNOW?

Peacocks are like living rainbows! the boys, called peafowls, show off their stunning feathers to impress everyone around them. these feathers have "eyes" that make them look like magical fans. Peacocks love to strut and dance, turning every day into a vibrant celebration.

COLOR TIP

For the peacock's feathers, use a mix of blues, greens, and golds. Capture the dazzling and iridescent colors that make peacocks extraordinary, adding layers to showcase the intricate patterns of their majestic plumage.

YAK

- Class: Mammalia
- Order: Artiodactyla
- Family: Bovidae
- Origin: Asia

DID YOU KNOW?

Yaks are the fuzzy superheroes of the mountains! With long, shaggy fur and big, strong bodies, yaks are built for the chilly Himalayan heights. They're not just good at carrying loads; they also provide milk, meat, and warm wool. Yaks are like the friendly giants of the snowy peaks!

COLOR TIP

For the yak's fur, use shades of browns, blacks, and grays. Capture the thick and woolly appearance of their fur, adding layers to showcase the rugged and resilient nature of these magnificent mountain-dwellers.

GERMAN SHEPHERD

- Class: Mammalia
- Order: Carnivora
- Family: Canidae
- Origin: Germany

DID YOU KNOW?

German Shepherds are the heroes of the canine world! they're not just known for their striking looks, with pointy ears and a bushy tail, but also for being super intelligent and loyal. these dogs are like detectives, often working with police and rescue teams. German Shepherds make loving family members and brave protectors.

COLOR TIP

For the German Shepherd's fur, use a mix of blacks, browns, and tan. Capture the sleek and powerful appearance of these incredible dogs, adding depth to their coat to showcase their intelligence and strength.

WILD BOAR

- Class: Mammalia
- Order: Artiodactyla
- Family: Suidae
- Origin: Europe, Asia, and North Africa.

DID YOU KNOW?

Wild boars are like forest explorers with a wild side! These sturdy pigs have strong tusks and a rough coat that helps them thrive in woodlands. They're skilled foragers and love to root around for tasty treats. Wild boars are tough, independent, and always ready for an adventure in the great outdoors.

COLOR TIP

For the wild boar's fur, use a mix of browns and grays. Capture the earthy and natural tones that complement the forest setting, adding layers to showcase the wild and untamed beauty of these woodland adventurers.

CAT

- Class: Mammalia
- Order: Carnivora
- Family: Felidae
- Origin: Worldwide

DID YOU KNOW?

Cats are the perfect companions! they're independent, playful, and masters of graceful movements. With their keen senses, cats are excellent hunters. they love cozy spots for napping and have a language of meows to communicate. Cats are like furry friends that bring joy and warmth to our homes.

COLOR TIP

For the cat's fur, use a mix of colors depending on the cat's breed—commonly whites, blacks, browns, and oranges. Capture the sleek and elegant appearance of these delightful creatures, adding textures to showcase the soft and luxurious nature of their fur.

SILKWORM

- Class: Insecta
- Order: Lepidoptera
- Family: Bombycidae
- Origin: China

DID YOU KNOW?

Silkworms are like tiny silk factories! They start as tiny eggs and spin delicate silk cocoons as they grow. Silkworms munch on mulberry leaves and create silky threads for their cocoons. Humans use these threads to make beautiful silk fabric. Silkworms are not just insects; they're little artists weaving a silky masterpiece.

COLOR TIP

For the silkworm's body, use shades of whites, yellows, and greens. Capture the delicate and soft appearance of these remarkable insects, adding layers to showcase the intricate patterns of their silk-producing journey.

PARTRIDGE

- Class: Birds
- Order: Galliformes
- Family: Phasianidae
- Origin: Worldwide

DID YOU KNOW?

Partridges are like nature's camouflaged gems! these ground-dwelling birds have intricate patterns on their feathers, helping them blend into their surroundings. Partridges are skilled at flying but prefer to run to escape danger. they're social birds and often seen in small groups, making the landscape even more vibrant.

COLOR TIP

For the partridge 's feathers, use a mix of earthy tones like browns and grays. Add subtle patterns to capture the intricate and natural camouflage that makes partridges stand out in their diverse habitats.

MOUSE

- Class: Mammalia
- Order: Rodentia
- Family: Muridae
- Origin: Worldwide

DID YOU KNOW?

Mice are like little acrobats of the animal world! They're quick, nimble, and can squeeze into tiny spaces. Mice have big ears to listen carefully, and they're excellent at nibbling on tasty treats. These tiny friends are social creatures, often seen in groups, and bring charm to their surroundings.

COLOR TIP

For the mouse's fur, use a mix of grays, browns, and whites. Capture the small and endearing appearance of these agile rodents, adding textures to showcase the soft and sleek nature of their fur.

RABBIT

- Class: Mammalia
- Order: Lagomorpha
- Family: Leporidae
- Origin: Worldwide

DID YOU KNOW?

Rabbits are known for their incredible agility and powerful hind legs. They are prolific breeders and have a keen sense of smell to detect predators. Rabbits are also expressive communicators, using thumps, grunts, and various body movements to convey messages to their fellow rabbits.

COLOR TIP

For the rabbit's fur, use a mix of whites, grays, and browns. Their coats can vary in color, so feel free to add different shades to capture the charming diversity of these hoppy little friends.

Thank you for making it this far!

I deeply appreciate the time you've dedicated to enjoying this book. As a newly publisher founded in 2023, it's a genuine pleasure and honor for me to share this wild journey with you or your children.

Therefore, if you have a minute, I would love to read your impression of this book on Amazon.

I want to know what you take away from it! :)

How to leave your review:

1. Open your camera on your smartphone.
2. Point your mobile phone camera at this QR code.
3. The webpage to write the review will appear in your browser.

Or visit website **tomeudf.com/reviewfarm** in your browser.

Made in United States
Troutdale, OR
04/12/2024